T0097364

# Energy-Sector Workforce Development in Southwestern Pennsylvania

Aligning Education and Training with Innovation and Needed Skills

Gabriella C. Gonzalez, Reema Singh, Rita Karam, David S. Ortiz

For more information on this publication, visit www.rand.org/t/rr807

Library of Congress Control Number: 2014955486

ISBN: 978-0-8330-8810-9

# Preface

In response to economic and environmental challenges, the U.S. energy sector has become highly innovative in developing and applying new technologies and procedures. These innovations often require more–highly skilled labor than was needed in the past. Because of industry demand for new skills, training and educational institutions need to anticipate technological advances to train the workforce in new competencies. To accommodate rapid changes in technology, community colleges and career and technology centers need to effectively and efficiently modify existing curricula and programs and develop new ones. In addition to learning core technical skills, workers need to acquire soft skills, such as professionalism, critical thinking, and problem-solving, to better adapt to unexpected changes in their work environments.

The National Energy Technology Laboratory asked the RAND Corporation to explore how innovation in the energy sector affects workforce needs in southwestern Pennsylvania. To perform this analysis, the RAND team worked closely with the Allegheny Conference on Community Development and the Three Rivers Workforce Investment Board. This report communicates the results of the study. It sets the stage for more–in-depth work supporting the development of programs to train workers in the energy sector. The results should be of interest to persons developing training programs to meet the needs of the energy sector and to firms in the energy sector, which need to plan for their future workforce needs.

## The RAND Environment, Energy, and Economic Development Program

The research reported here was conducted in the RAND Environment, Energy, and Economic Development Program, which addresses topics relating to environmental quality and regulation, water and energy resources and systems, climate, natural hazards and disasters, and economic development, both domestically and internationally. Program research is supported by government agencies, foundations, and the private sector.

This program is part of RAND Justice, Infrastructure, and Environment, a division of the RAND Corporation dedicated to improving policy and decisionmaking in a wide range of policy domains, including civil and criminal justice, infrastructure protection and homeland security, transportation and energy policy, and environmental and natural resource policy.

Questions or comments about this report should be sent to the project leader, Gabriella C. Gonzalez (Gabriella_Gonzalez@rand.org). For more information about the Environment, Energy, and Economic Development Program, see http://www.rand.org/energy or contact the director at eeed@rand.org.

# Contents

# Figures and Tables

## Figures

## Tables

# Summary

## Background and Study Objectives

Energy resources have defined the southwestern Pennsylvania (SWPA) region, the 32-county region surrounding the city of Pittsburgh, stretching across western Maryland, eastern Ohio, southwestern Pennsylvania, and the northern panhandle and north-central regions of West Virginia. The region's abundant coal resources fueled the U.S. steel industry for many decades; oil was first produced commercially there, and the region gave birth to the U.S. oil industry. Natural gas, first used to evaporate water to produce salt, is once again remaking the area. Yet, the energy industry is broader than these three fossil resources. It includes firms that design and manufacture the specialized equipment used to seek, extract, and profitably use these resources. The energy industry also includes nuclear, wind, and solar energy and the technologies to integrate them, such as power electronics and automatic controls that conserve energy. A workforce has developed in SWPA to support these industries.

The energy sector is a vital component of the region's economy. In the ten-county region surrounding the city of Pittsburgh, the energy sector constitutes 16 percent of the region's economic activity and generates $19 billion in gross regional product. According to 2012 analyses conducted by the Pittsburgh Regional Alliance using Economic Modeling Specialists International data, nearly 36,000 people in the region worked in the energy sector at 981 establishments that are directly or indirectly related to the extraction, production, and use of energy resources. (This number includes corporate employees and the self-employed.) Since 2007, the energy sector has experienced a 23.4-percent increase in employment, which can be directly attributable to the concomitant increase in natural-gas extraction (50.6 percent) that occurred in the same period in the Marcellus Shale (Pittsburgh Regional Alliance, 2013).

The growth in the energy sector and the fact that major federal research energy laboratories and coal-development centers are located in the area have driven technological innovations and the corresponding need for workers at all skill levels, including those with trade certifications, associated degrees, and four-year college degrees (Kauffman and Fisher, 2012). Technological advancements across the energy sector require that the sector continuously innovate. This innovation includes "learning-by-doing" improvements in resource extraction exemplified by oil and gas industries, the advances in materials and manufacturing that reduce costs and improve the performance of solar-photovoltaic cells, new power electronics that will enable active monitoring and control of bulk electricity, and innovative system designs for nuclear-power plants. The scale and pace of these innovations will require workers throughout the energy sector and the energy supply chain and manufacturing capacity to add to their existing competencies, as well as develop new skill sets. Technological innovation also tends to reduce demand for lower-skill labor while increasing demand for higher-skill labor.

To address the challenges of ensuring a skilled, adaptable workforce for the energy sector, the National Energy Technology Laboratory asked RAND to partner with the Allegheny Conference on Community Development and the Three Rivers Workforce Investment Board to help determine how the postsecondary education and training system (which includes short-term certification and workforce-training programs, community-college programs conferring associate's degrees, and four-year bachelor's degree–granting institutions) could meet the growing and shifting skill demands for labor from the energy sector due to technological innovation. Our study focused on the shifting demands on the semiskilled workforce in SWPA from now through 2020.

The study had three objectives:

1. Document key technological innovations currently taking place in the energy sector as a way to better predict where growth in jobs and shifts in skills may be needed
2. Identify possible best practices of educational and training programs that have successfully responded to innovations in other sectors
3. Determine the implications of these innovations and provide recommendations for the energy-sector education and training system in SWPA.

The findings from this study can be used as a springboard for deeper discussion among regional education and training institutions, business leaders in the energy industry, and nonprofit organizations devoted to supporting the employability of local talent in the energy sector. We expect that this report will be used in a workshop setting in which these regional partners initiate specific practices to collaboratively develop programming, policies, and implementation strategies to ensure that administrators of postsecondary education and training institutions undertake practices to best prepare programs that support employment of local talent in the energy sector.

## Analytic Approach

To accomplish the first objective, RAND researchers turned to experts in the field, as well as national research publications, to identify key technological innovations currently taking place in the various fields making up the energy sector, in which we include coal, natural gas, nuclear power, solar energy, wind, electricity transmission and distribution, and intelligent building technologies. To gain further insights into local trends and to help determine which innovations might be implemented in the near term and whether these changes would entail incremental advances or wholesale changes, we also discussed technological developments within those fields with regional and national academic and professional experts.

For the second objective, we reviewed the literature on U.S. and international education and training systems to determine key features of those systems that have successfully responded to technological innovations in regional industries. We identified five characteristics of education and training systems that lead to success in meeting the changing needs of labor markets. We determined the relevance and application, if any, of these characteristics to the Pittsburgh region by speaking with local directors of four representative education and training institutions in the region that provide certifications, associate's degrees, or bachelor's degrees in fields intended to lead to employment in the energy sector. Interviews provided insights on the strengths of these programs, possible areas in need of improvement, training providers' perspectives on where gaps in training and employment may lie, and training providers' perspectives on their programs' responsiveness to innovation in the workplace. Because of the small number of education and training institutions included in this study, findings are not intended to be generalizable. Instead, they offer suggestive insights into how education and training is conducted in the region.

For the third objective, we synthesized the findings from the analyses conducted in the first two tasks to craft measures the region's energy-sector stakeholders could employ to support the education and training of regional talent. We intend for these suggested measures to spark discussion among representatives from the region's energy-sector businesses and postsecondary educational institutions.

## Key Findings

### Four Drivers Motivate Innovation in the Energy Sector

Our focused literature review and interviews revealed that, although the energy sector increasingly relies on technological innovation, innovation tends to take place on a small scale and incrementally. We found that innovation in the energy sector that affects necessary workforce skills can be organized into the following motivators:

- increasing productivity in energy extraction
- minimizing risk of environmental damage and reducing emissions of carbon dioxide
- integrating renewable energy into the grid
- improving energy end-use efficiency.

### The Energy Sector Needs an Agile, Skilled Workforce to Adapt to Changes in Technology and Innovation

National and regional experts with whom we spoke consistently identified core competencies needed for the energy workforce to respond to technological innovations. These included the following:

- content knowledge of the industry in question
- transferable, hands-on skills, including facility with computers and information technology (such as data analytics and geographic information system mapping), welding, and other practical skills workers can use across industries
- soft skills or workplace competencies, such as dependability, safety awareness, and decisionmaking and problem-solving.

### Five Practices Characterize Postsecondary Training and Education Programs That Best Meet the Needs of an Evolving Labor Market

Our literature review revealed that postsecondary education and training programs that tend to be the most successful in adapting to evolving labor-market needs are those that tend to have the following characteristics:

- **a well-developed mechanism to anticipate demand for specific occupations by developing consistent, ongoing relationships with industry leaders.** Understanding and documenting which occupations are in most employer demand is vital to ensuring that an education and training institution is relevant and produces graduates who are employable. This can occur through two approaches:

- systematically involving employers and social partners in setting policies, training content, and qualifications
- engaging in ongoing research activities and surveys of employers to assess demand for skills and quality of graduates (Organisation for Economic Co-operation and Development [OECD], 2010).

- **curricula that include occupation-specific and generic skills.** The curricula of career- and technical-training programs need to address the development of occupation-specific skills, as well as broad (generic) skills. Curricula should also address the development of soft skills, such as the ability to work in teams, communication skills, entrepreneurship, customer relations, and work discipline (OECD, 2010). Studies have shown that integrity and maturity are the skills most lacking among employees entering the workforce directly out of high school, followed by basic reading, writing, and social skills (OECD, 2010).

- **complementary in-class and workplace learning opportunities.** There is consensus in the literature about the importance of balancing educational (in-class) training with workplace (on-the-job) training. The educational setting is seen to be effective in developing literacy and numeracy, as well as developing knowledge in vocational theory (OECD, 2010). Workplace learning, on the other hand, is also considered important because it can solidify classroom concepts, promote the development of technical skills on modern equipment, and grow soft skills through real work experience (Aarkrog, 2005; Lasonen, 2005).

- **quality instructors with workplace and teaching experience.** The quality of the teaching and training profession is critical to effective learning in career- and technical-education programs. Instructors should have both pedagogical skills and work experience. Continuing education and professional development or training need to be provided to help instructors upgrade their skills to keep pace with changes in the occupations and technologies used.

- **ongoing quality-assurance processes.** Career- and technical-education programs need to engage in quality assurance. Quality assurance can range from external evaluation and accreditation to self-evaluation to regulation through professional associations.

## Southwestern Pennsylvania's Regional Education and Training Providers Display Some Aspects of the Five Promising Practices

- **Mechanisms for anticipating occupation demands through relationships with industry leaders were not typical.** The majority of the training programs and community colleges worked with advisory committees to help them develop the curricula and anticipate changes in occupation demand and skill sets in the energy sector. However, the frequency of interaction with the advisory boards tended to vary among the training programs: One training program met with its advisory board only every three years. Further, some programs relied on employers to initiate contact with them to inform the programs about changes needed to their curricula, rather than taking the initiative or having a systematic, ongoing, reliable process to obtain this information.

- **Curricula across the institutions included occupation-specific and generic skills, but delivery varied.** All education and training programs included in our study had curricula that included the occupation-specific content knowledge for the degree or certification. And most training programs and community colleges in our study indicated that their curricula incorporate generic and soft skills in addition to the occupation-specific content knowledge. Business partners, contractors, and employers with whom we spoke substantiated this claim. However, there was variation in how these skills were delivered: Some programs dedicated time to the development of these skills; others required instructors to integrate soft skills into their regular courses. In the latter case, interviewees did not know the extent to which instructors addressed the development of generic skills.

- **Workplace learning opportunities to complement in-class opportunities were not evident.** All of the education and training programs included in our study incorporated a blended approach to learning that included classroom instruction and laboratory experience. None of the respondents at each program mentioned having students develop work experience through internships at a workplace or other opportunities for on-the-job experience.

- **Hiring quality instructors with workplace and teaching experience has proven difficult.** Education and training programs included in our study had limited success in hiring trainers and instructors with direct experience in the industry. The study institutions tend to provide professional development to those instructors they hire from industry to improve their pedagogical skills. Education and training programs that aim to provide certifications and degrees that can lead to employment in the energy sector need to strengthen professional-development opportunities available to their instructors.

- **Programs need to undertake ongoing quality-assurance practices.** All the education and training programs in this study engage in some type of quality-assurance process but vary in the strategies they take. Some utilize elaborate, ongoing external and internal evaluation strategies, while others participate in much more simplified processes.

## Suggested Measures

Informed by our analysis of the above sources of information, we recommend the following to promote postsecondary education and training providers' abilities to successfully train talent in the skills needed in the energy sector in the Pittsburgh region. For these measures to be successful in supporting the energy-sector employment pipeline, educational institutions and employers need to consider a regional approach. The measures listed here are intended to be a springboard for further discussion among key stakeholders in the region.

## Develop Sustained and Continuous Partnerships Between Industry Leaders and Education and Training Providers

Employers in the energy sector should engage with regional education and training providers—such as local community colleges, unions, and private certification and training institutions—to forecast demand, provide regular input on curricula, support the acquisition of equipment and supplies used in hands-on training, and institute formalized workplace learning opportunities for students. There are various ways to develop a sustained partnership between industry leaders and training providers. First, industry leaders can serve on advisory boards or committees of training institutions and community colleges to have an active role in the development of curricula that meets the industries' needs. Another vehicle for partnership is establishing internship programs. Having such joint programs between employers and training institutions will naturally facilitate ongoing conversations because the parties would need to communicate regarding the structure and the content of the internships. Last, regional and local forums made up of leading companies and education and training institutions can be established to encourage ongoing communication regarding innovations, changes in the energy sector, and future job opportunities and related skill needs.

## Develop Education and Training Programs That Integrate Technical, Occupation-Specific Training Along with Workplace Readiness and Other Soft Skills

Incorporating soft skills in a program's curriculum will help prepare graduates for on-the-job demands. Integrating technical training that conveys specific content knowledge along with work readiness and basic skill training (coupled with individual case-management services) allows students in the training programs to obtain the wide range of content and skills necessary to be competent, yet agile, employees who are currently in demand. Including formalized internships or apprenticeships with partner employers will support this.

## Encourage Recruitment and Retention of Quality Instructors

We suggest two approaches to ameliorate the challenge of hiring and retaining quality instructors:

- Institute continuous professional development for current instructors.
- Develop agreements between training institutions and energy-sector employers in the region so that employees can teach in education and training programs and so that nonemployee instructors have the opportunity to gain workplace knowledge as part of their professional development.

## Incorporate Mechanisms to Continuously Ensure Quality of Education and Training Programs

In this study, all education and training programs that we examined were accredited. However, accreditation checks usually occur every few years to ensure that institutions comply with standards. Given the pace and scope of innovation and changing labor-market needs in the energy sector, education and training providers would benefit from having ongoing input to monitor whether their programs are meeting industry needs.

To complement accreditation, we suggest that education and training providers in the region develop and implement a mechanism that provides *continuous* and *rigorous* feedback. Two possible strategies are

- oversight by an external body that includes industry leaders
- internal assessment and ongoing monitoring.

## Document Progress on Whether Regional Energy-Sector Training and Employment Goals Are Being Met

To measure successes, it is important to document which efforts or initiatives are working to meet the region's goals in hiring, retention, and employment of local talent in the energy sector. We suggest that industry leaders and education and training providers collaborate on reporting to the community on at least an annual basis about the efforts that are working, lessons learned, and plans for further collaboration in the coming year. To make this type of reporting possible, we recommend the development of a monitoring system that captures information on the quality and characteristics of the various training programs or efforts in the region and that collects information on measurable outcomes, such as student graduation rates, job placement (e.g., the types of industries and companies in which graduates were hired), retention, and employer satisfaction with the quality and performance of the programs' graduates. The design of such a system and consensus on the type of information that should be included will require input from various stakeholders, including education and training programs in the region.

# Acknowledgments

The RAND team is deeply indebted to many people and institutions throughout southwestern Pennsylvania that helped us to scope and complete the work. Laura Fisher, senior vice president for special projects at the Allegheny Conference on Community Development, helped us to define the study questions and facilitated our interactions with many firms and programs in the region. Stefani Pashman, chief executive officer of the Three Rivers Workforce Investment Board, was a core partner, providing key insights regarding the needs of employers and the capabilities of the regional workforce. We appreciate the participation of business and community leaders in a workshop that we held in April 2013 to initiate the project. We are also grateful for the time and insights provided by our study participants (regional and national experts, as well as key staff from the four education and training programs we interviewed). The National Energy Technology Laboratory provided financial sponsorship for the project and supported our partnerships throughout the region. The late Anthony Cugini, former director of the laboratory; Frederick Brown, Gregory Kawalkin, and Thomas J. Feeley III were all instrumental in initiating the study. We would also like to thank the careful reviews of Keith Crane, Tom Witt, and Robert Bozick. Any errors and omissions are the fault of the authors.

# Abbreviations

| | |
|---|---|
| AD | associate's degree; degree completed after two years of full-time schooling beyond high school |
| BD | bachelor's degree; degree completed after four years of full-time schooling beyond high school |
| BD+ | bachelor's degree; degree completed after four years of full-time schooling beyond high school with some period of related work experience required |
| $CO_2$ | carbon dioxide |
| DAYCUM | design and curriculum |
| EMSI | Economic Modeling Specialists International |
| GHG | greenhouse gas |
| HPO | high-priority occupation |
| LT-OJT | long-term on-the-job training; high-school diploma and at least one year of on-the-job training or an apprenticeship |
| MT-OJT | medium-term on-the-job training; basic tasks and skills are learned through a period of on-the job training, and a high-school diploma may be required |
| OECD | Organisation for Economic Co-operation and Development |

| | |
|---|---|
| PV | photovoltaic |
| SWPA | southwestern Pennsylvania |
| TVET | technical and vocational education and training |
| WK EXP | work experience; a high-school diploma and training gained through hands-on work in a similar occupation |

# CHAPTER ONE

# Aligning Workforce Education and Training with Energy-Sector Needs

## Purpose of This Study

To address the challenges of ensuring a skilled, adaptable workforce for the energy sector, the National Energy Technology Laboratory (NETL) asked the RAND Corporation to partner with the Allegheny Conference on Community Development and the Three Rivers Workforce Investment Board (TRWIB) to examine the role of technological innovation in transforming and defining the competencies of workers needed to fill jobs in high demand in the energy sector in southwestern Pennsylvania (SWPA) from now through 2020.[1]

We define SWPA as those 32 counties identified as part of the Power of 32,[2] which include western Maryland, eastern Ohio, southwestern Pennsylvania, and the northern panhandle and north-central regions of West Virginia, as illustrated in Figure 1.1.

This project focused on semiskilled jobs—those that require an employee to have at least a high-school degree plus a certification, specialized training, or an associate's degree in the appropriate field—as well as some jobs that require bachelor's degrees. The project explored how postsecondary education and training institutions in the region could be strengthened to be responsive to any new or emerging workforce demands. Incremental advances in existing technologies may require fewer curriculum changes or certificate-requirement modifi-

---

[1]  As identified Kauffman and Fisher, 2012.

[2]  The Power of 32 is a project of multiple foundations, for-profit corporations, and providers of postsecondary education. For more information, see Power of 32, undated (b).

**Figure 1.1**
**The 32 Counties That Make Up the Southwestern Pennsylvania Region**

SOURCE: Power of 32, undated (a).
RAND RR807-1.1

cations on behalf of the education and training institutions but may require more creative skills at the job site.

This study had three objectives:

- Document near-term trends in innovation and technology to determine areas in which the energy sector is innovating and, therefore, where growth in jobs is likely and where shifts in skills are needed.
- Identify features of educational and training programs that have successfully responded to innovations in other sectors.

- Extract key implications for workforce-development programs focused on the energy sector in SWPA.

## Background

The extraction, production, and distribution of energy to be used for residential and commercial consumption is a fundamental part of the U.S. economy: The United States is the world's largest generator of electricity, with about 1,100 gigawatts of generating capacity. According to recent estimates from a variety of sources, nationwide, the energy industry employed approximately 817,498 people in oil and gas jobs in 2010; 120,000 in nuclear-power plants in 2010; between 170,000 and 392,000 in mining operations in 2008; 119,000 jobs in solar in 2012; 75,000 in wind-related jobs in 2011; and 5,200 in geothermal jobs in 2010 (Committee on Emerging Workforce Trends in the U.S. Energy and Mining Industries et al., 2013, pp. 1–4, 23, 52, 68, 106, 120, 143). See Figure 1.2.

Although the state of Pennsylvania is only one portion of the region that we cover in this report, it serves as a strong example of regional trends in the energy sector and in the workforce. Pennsylvania is the second-largest generator of electricity (producing more than 137 million megawatt-hours) in the nation, accounting for 6 percent of the nation's total. The state is also the largest net exporter of electricity of any state, exporting approximately 35 percent of the 228 million megawatt-hours generated in 2011. The sources for this electricity have shifted through the years. In 2000, coal was responsible for 57.6 percent of the electricity generated (natural gas accounted for 1.3 percent). By 2010, the shares of natural gas and renewables had grown substantially: In 2010, natural gas accounted for 14.7 percent of electricity generation, and the share of coal had fallen to about 50 percent.[3]

Table 1.1 itemizes the number of direct jobs (those that are created specifically by businesses that produce, extract, transport, or

---

[3] All information in this paragraph comes from the U.S. Energy Information Administration, 2014.

**Figure 1.2**
**Nationwide Energy-Sector Jobs**

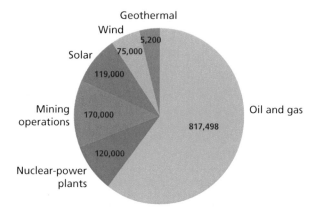

SOURCE: Committee on Emerging Workforce Trends in the
U.S. Energy and Mining Industries et al., 2013, pp. 4, 23, 52,
68, 106, 120, and 143.
NOTE: Data for oil and gas, nuclear-power plants, and
geothermal are from 2010. Data for mining operations are
from 2008; the number shown is the low-end estimate, and
the high-end estimate was 392,000. Data for wind are from
2011; data for solar are from 2012.
**RAND** *RR807-1.2*

**Table 1.1**
**Direct and Indirect Employment in the Energy Sector in Pennsylvania**

| Industry | Coal (2010) | Electric Power (2010) | Natural Gas and Crude Oil (2012) |
|---|---|---|---|
| Direct jobs | 20,700 | 16,532 | 25,628 |
| Indirect and induced jobs | 42,290 | 53,384 | 77,040 |
| Total | 62,990 | 69,916 | 102,668 |

SOURCES: Coal-mining employment figures are from National Mining Association,
undated (cited in Commonwealth Economics, 2013, Table 2-11, p. 47); electric
power–production figures are from Edison Electric Institute (cited in Commonwealth
Economics, 2013, Table 3-16, p. 71); natural-gas and crude-oil figures from IHS Global
Insight (now IHS Economics and Country Risk) (cited in Commonwealth Economics,
2013, Table 2-15, p. 52).

refine energy) and indirect, or induced, jobs (those that are peripherally created around the businesses) that have been attributed to the electric-power, coal-mining, natural-gas, and crude-oil industries in Pennsylvania. Direct jobs in the coal, natural-gas, and crude-oil industries include mine or drilling work, support activities, and transportation. Direct jobs in the electric-power industry include electric-power generation, transmission, and distribution; indirect jobs are primarily suppliers.

The Pennsylvania Department of Labor and Industry has identified high-priority occupations (HPOs) in the state. HPOs meet three criteria:

1. They are in demand by employers.
2. They have higher skill needs.
3. They provide family-sustaining wages.[4]

Table 1.2 lists the HPOs relevant to the energy sector in the state. This table notes all HPOs within the energy and advanced manu-

**Table 1.2**
**High-Priority Occupations in Energy-Related Industries in Pennsylvania, 2013**

| Occupation | Annual Openings | Required Education | Industry |
|---|---|---|---|
| Heavy or tractor-trailer truck driver | 2,575 | WK EXP | Energy |
| Automotive service technician or mechanic | 1,258 | LT-OJT | Advanced manufacturing |

---

[4] The Pennsylvania Department of Labor and Industry classifies occupations as HPOs through an analysis of the Pennsylvania targeted-industry clusters in three steps:

- Identify occupations that are important to the success of industries within the clusters.
- Review occupational statistics, and apply filters to hone the list.
- Incorporate input from industry experts and regional leaders.

More information about data and calculations to determine which occupations are considered to be high-priority can be found at Pennsylvania Department of Labor and Industry, undated.

**Table 1.2—Continued**

| Occupation | Annual Openings | Required Education | Industry |
|---|---|---|---|
| Maintenance or repair worker (general) | 1,443 | MT-OJT | Advanced manufacturing, energy |
| Operating engineer or other construction-equipment operator | 828 | MT-OJT | Energy |
| Production, planning, or expediting clerk or purchasing agent | 768 | MT-OJT or LT-OJT | Advanced manufacturing |
| Industrial machinery mechanic | 738 | LT-OJT | Advanced manufacturing |
| Supervisor for construction trades and extraction workers | 675 | WK EXP | Energy |
| Inspector, tester, sorter, sampler, or weigher | 593 | MT-OJT | Advanced manufacturing |
| Welder, cutter, solderer, or brazer | 528 | MT-OJT | Advanced manufacturing |
| Machinist | 508 | LT-OJT | Advanced manufacturing |
| Mechanical engineer | 452 | BD | Advanced manufacturing |
| Supervisor for mechanics, installers, and repairers | 448 | WK EXP | Advanced manufacturing, energy |
| Parts salesperson | 411 | MT-OJT | Advanced manufacturing |
| Industrial engineer | 359 | BD | Advanced manufacturing |
| Bus or truck mechanic or diesel-engine specialist | 353 | LT-OJT | Energy (logistics and transportation) |
| Computer-controlled machine tool operator (metal or plastic) | 249 | MT-OJT | Advanced manufacturing |
| Electrical power-line installer or repairer | 264 | MT-OJT | Energy |

**Table 1.2—Continued**

| Occupation | Annual Openings | Required Education | Industry |
|---|---|---|---|
| Industrial production manager | 217 | BD+ | Advanced manufacturing |
| Extruding or drawing machine setter, operator, or tender (metal or plastic) | 207 | MT-OJT | Advanced manufacturing |
| Heat-treating equipment setter, operator, or tender (metal or plastic) | 138 | MT-OJT | Advanced manufacturing |
| Multiple machine tool setter, operator, or tender (metal or plastic) | 119 | MT-OJT | Advanced manufacturing |
| Rotary-drill operator (oil or gas) | 134 | MT-OJT | Energy |
| Service-unit operator (oil, gas, or mining) | 131 | MT-OJT | Energy |
| Wellhead pumper | 114 | MT-OJT | Energy |
| Industrial engineering technician | 98 | AD | Advanced manufacturing |
| Maintenance worker (machinery) | 71 | MT-OJT | Advanced manufacturing |
| Computer–numerically controlled machine tool programmer (metal or plastic) | 39 | MT-OJT | Advanced manufacturing |

SOURCE: Pennsylvania Department of Labor and Industry, undated.

NOTE: WK EXP = work experience; a high-school diploma and training gained through hands-on work in a similar occupation. LT-OJT = long-term on-the-job training; high-school diploma and at least one year of on-the-job training or an apprenticeship. MT-OJT = medium-term on-the-job training; basic tasks and skills are learned through a period of on-the job training, and a high-school diploma may be required. BD = bachelor's degree; degree completed after four years of full-time schooling beyond high school. BD+ = BD with some period of related work experience required. AD = associate's degree; degree completed after two years of full-time schooling beyond high school.

facturing industries in Pennsylvania, the number of openings, and required educational attainment. Of note is that only a few of these HPOs require a college education. In fact, most require a high-school education with some on-the-job training or some work experience.

These are noted in the "Required Education" column in the table as AD (degree completed after two years of full-time schooling beyond high school), LT-OJT (high-school diploma and at least one year of on-the-job training or an apprenticeship), MT-OJT (medium-term training in which a high-school diploma may be required), and WK EXP (a high-school diploma and training gained through hands-on work in a similar occupation).

The geographic core of SWPA, the ten-county area directly surrounding the city of Pittsburgh, experienced large increases in energy-related job openings between 2006 and 2011. Overall employment in the energy sector (related to extraction, generation, and supply chain) increased by 6.3 percent between 2007 and 2011, compared with a 2.4-percent decrease in overall employment in the region. Eighty-eight percent of these job openings were in the natural-gas and coal energy fields, while 64 percent of the announced expansions were in manufacturing (there could have been overlap with the natural-gas and coal energy fields) (Kauffman and Fisher, 2012). There was also job expansion, although to a lesser extent, in energy-related businesses, such as transportation and distribution.

Analysis conducted by the Allegheny Conference's Pittsburgh Regional Alliance (2013) using data from Economic Modeling Specialists International (EMSI) found that 35,980 people worked at 981 establishments in the energy sector in the ten-county region in 2012. Employment includes corporate employees in addition to self-employed individuals. EMSI data reveal that, between 2007 and 2012, the energy sector in the ten-county region experienced a 23.7-percent increase in employment, primarily due to the growth of jobs in extraction in the Marcellus Shale. Table 1.3 lists the key employers in the ten-county region.

Two factors are pressuring postsecondary education and training institutions to produce a highly skilled and agile workforce to keep pace with the sector's evolving requirements. First, projected demand for workers has increased—from both established energy industries, such as oil and gas extraction and coal mining, and renewables—due to increased production and extraction, coupled with impending retirements from the baby-boomer generation. More jobs will be opening up

Table 1.3
**Major Employers, Estimated Employment, and Services in the Ten-County Pittsburgh Region, 2012**

| Company | Estimated Employment | Product/Service |
|---|---|---|
| Westinghouse Electric Company | 5,600 | Nuclear-power development |
| Consol Energy | 3,503 | Coal, natural gas |
| FirstEnergy Corporation | 3,000 | Electric-power generation |
| Bechtel Group | 3,000 | Nuclear-power research, components |
| Alpha Natural Resources | 1,800 | Coal, natural gas |
| Siemens AG | 1,732 | Electrical component manufacturing |
| Eaton Electrical Sector | 1,286 | Electrical component manufacturing |
| Mine Safety Appliances | 1,275 | Mining equipment |
| Duquesne Light Holdings | 1,200 | Electric-power distribution |
| Rosebud Mining | 1,140 | Coal |
| Elliott Group | 1,100 | Compressor manufacturing |
| EQT Corporation | 1,015 | Natural gas |
| Curtiss-Wright Corporation | 1,000 | Electrical component manufacturing |
| Peoples Natural Gas | 976 | Natural-gas distribution |
| Mitsubishi Electric Corporation | 650 | Electrical component manufacturing |
| Emerson Process Management | 600 | Electrical component manufacturing |
| Direct Energy | 600 | Energy distribution |
| Chevron Corporation | 580 | Natural gas |
| Chesapeake Energy Corporation | 500 | Natural gas |
| WESCO International | 500 | Electrical component distribution |
| Caterpillar | 500 | Mining equipment |

SOURCE: Pittsburgh Regional Alliance, 2013.

in the near term than the current output of graduates is likely to meet. A 2011 Center for Energy Workforce Development survey of energy-sector employers found that, by 2015, 36 percent of positions considered critical by the industry (i.e., skilled utility technicians and engineering, excluding positions in nuclear) may need to be replaced because of potential retirement or attrition, with an additional 16 percent needing to be replaced by 2020. This accounts for nearly 110,000 employees in positions that the industry identifies as the most critical (Center for Energy Workforce Development, 2012, p. 3). Second, the U.S. energy sector has become highly innovative in developing and applying new technologies and procedures, particularly in renewables, mining, and extracting natural gas (Committee on Emerging Workforce Trends in the U.S. Energy and Mining Industries et al., 2013). Because these innovations often require more–highly skilled labor to manufacture and deploy energy technologies and systems, many members of the workforce may need to improve their skills. Education and training institutions therefore need to be well equipped to create program-ming and curricula that anticipate technological advances to train (and retrain) the workforce in new competencies.

In the SWPA region, these issues are even more pronounced than in the United States as a whole, given the burgeoning field of natural-gas extraction that exists alongside the well-established, traditional coal-mining, crude-oil, and nuclear-power industries. Nowadays, there is an increasing demand for workers at all skill levels, ranging from trade certification to associate's degree to graduate degree. The occupa-tions in most demand are those that require only some training or cer-tification after high school (Kauffman and Fisher, 2012). In addition, because of technological innovation, the type of skills that workers need to develop may be different from those needed previously. How-ever, there has been documentation that the workforce-development system in the Pittsburgh region is fractured and not well aligned with workforce demands (*Western Pennsylvania's Workforce Development System*, undated).

## Analytic Approach

For the analyses undertaken in this study, we defined the energy sector broadly to include the seven industries listed in the Kauffman and Fisher (2012) report: oil, natural gas (including liquefied natural gas [LNG] and natural-gas liquids), coal, renewables (e.g., solar and wind), nuclear, electricity transmission and distribution, and intelligent building technologies. Within those industries, we focus on a subset of energy-industry components:

- production and extraction
- equipment manufacturing
- delivery and transport
- monitoring and controls
- design and construction (including smart buildings).

It is important to note that we did not investigate workplace needs or technological advances in downstream industries typically supported by the energy sector, such as refining of petroleum crude oil and the processing and purifying of raw natural gas, or the marketing and distribution of products derived from crude oil and natural gas. Nor did we include in this study indirect jobs that typically build up around newly established energy-sector locales, such as hotels or restaurants and other hospitality or service-sector businesses.

### Task 1: Examine Near-Term Innovation in the Energy Sector

To examine potential future directions of the energy sector in SWPA and the role particular innovations may have within it, RAND researchers first reviewed key research documents addressing energy innovation published since 2010 in academic journals and by national and federal organizations. This allowed us to gain a comprehensive overview of the kinds of technological innovations taking place within various energy sectors. In doing so, we were able to identify specific innovations being developed in the nuclear, coal, solar, wind, natural gas, electricity transmission and distribution, and intelligent building–technology sectors. Identifying these technologies enabled us to conceptualize trends and motivators of technological innovations, which helped inform discussion protocols.

We then conducted interviews in person and over the telephone with a purposefully selected representative sample of nine local industry business and nonprofit leaders and two entrepreneurs whose establishments are located within the 32-county footprint of the Power of 32 SWPA region. We also spoke with 12 academic researchers familiar with the key technologies, competitive pressures, and likely innovations in the energy sector. We selected the academic researchers we interviewed for their specialized expertise in several core fields across the energy sector. We conducted conversations with three experts in coal, three experts in natural gas, two experts in nuclear energy, two experts in solar energy and electricity, one expert in intelligent building technology, and one expert in wind energy. When speaking with business leaders in the energy industry, we conducted discussions with employers from companies both upstream (those involved in production and extraction of resources to create energy) and downstream (those involved in the refinement and transportation of energy) to understand how advances are affecting supply-chain processes as a whole. In total, we spoke with two leaders from upstream organizations and seven leaders from downstream organizations. We also interviewed one nonprofit leader to gain insights into the implications of technological innovation within the advocacy and nonprofit sectors.

Although we selected interviewees for this study to provide a representative and broad perspective, it is important to acknowledge that we did not engage with all energy-sector business and nonprofit leaders, nor with all academic or government researchers who are working on topics related to energy or innovation. This means we could have inadvertently not included a key perspective. Further, the sample could potentially be skewed to those academic researchers who are more established and published. Therefore, we could have missed the perspective of more-junior academic researchers new to the field.

## Task 2: Benchmark Educational and Training System Designs Addressing Innovation's Effects on Workforce Development

This task identified the features of educational and training systems that have documented successes in responding to technological innovation and changes in labor demand in a range of sectors both within the United States and internationally. We identified exemplary technical and vocational education and training (TVET) systems and reviewed them as in-depth case studies.

We then employed a case-study methodology of four exemplar educational and training programs located in the 32-county SWPA region that train workers for the regional energy sector, examining how their programs compare with the characteristics determined as model practices from the literature review on exemplary TVETs. This case-study approach allowed us to develop an in-depth understanding of the programming and curricula that each institution utilizes to support the education and training of students for employment in the energy sector. Case-study analysis can be susceptible to bias in the interpretation of findings because only a handful of sites are used to gather data. To overcome this potential bias, the RAND team developed semi-structured interview protocols to ensure that we addressed relevant topics and asked similar questions across the sites. Furthermore, we are cautious not to generalize our interpretation of the data beyond the four sites that participated in the study. In addition, we purposefully selected the four programs to serve as case studies to be representative of the varied types of education and training institutions that exist in the Power of 32 SWPA region. However, it is important to note that we do not intend for these institutions to be representative of *all* education and training providers. Nor do they offer a comprehensive understanding of experiences and perspectives of all staff or students at these institutions. Instead, they offer an understanding of each program director's specific experiences and perspectives. The example programs in our study sample included one community college, one institution of higher education that offers both bachelor's and associate's degrees, one union apprenticeship program, and one career- and technical-education center. We interviewed at least one program director from each institution. Interview questions inquired about each program's design features, perceived strengths of these programs in relation to the features, possible areas in need of improvement, training providers' perspectives on where gaps in training and employment could lie, and training providers' perspectives on their programs' responsiveness to innovation in the workplace.

**Task 3: Extract Key Implications for Southwestern Pennsylvania**
Task 3 synthesized the analyses conducted in the previous two tasks to develop a set of measures designed to improve the responsiveness of educational and training programs to changes in energy-sector skill needs resulting from technological innovations.

These suggested measures are intended to spur discussion among key education and industry leaders, human-resource professionals, and nonprofit stakeholders in the region to determine a feasible, practical, and actionable strategy to improve the alignment of education and training with industry needs in the energy sector.

## Organization of This Report

We have structured the remainder of this report according to the study's objectives. We first document trends in innovation and technology in the energy sector and the demand for labor and skills likely to be needed in the near term, as determined in our literature review and interviews with academic researchers and with energy-sector business and nonprofit leaders in the region. Second, we report the findings of our review of the literature on best practices in education and training, particularly in the energy sector and for semiskilled workforce training. Next, we report our findings from our exercise to benchmark education and training programs in SWPA to determine the extent to which these example programs compare with the promising practices synthesized from the literature review. We conclude the report with suggestions for specific actions applicable to SWPA to promote current or design new education and training programs to best meet the shifting demands for workers able to work with new technologies in the energy sector in the region. We intend the final section to be used to organize discussions with local stakeholders to assist in developing a workforce-development strategy for the region.

# CHAPTER TWO

# Skills Needed to Meet Changing Industry Demand

In this chapter, we document areas in which the energy sector is growing and innovating, areas that affect job growth and the demand for skills. We look at the present and the near term across the seven industries listed in Chapter One. To determine what innovations might be implemented in the near term, examining whether anticipated innovations are revolutionary or incremental advances on existing technologies and their likely implications for demand for labor, we document the competitive pressures and technological solutions that the energy sector faces. For the information in this section, we relied on our review of the literature; interviews with key business leaders, entrepreneurs, and researchers in the energy field; and statistics on changes in jobs.

## Technological Innovation in the Energy Sector

A key finding from our literature review and interviews is that the energy sector relies heavily on technology and innovation: Increasing efficiency and lowering costs of extraction, production, or delivery are vital to the success of businesses in the sector. Shifts in technologies continue to change demands on workers. For example, production companies need the ability to analyze vast amounts of data to find new locations to extract oil or gas; robots are now used to repair components of nuclear plants; and extraction companies employ highly sophisticated drilling equipment. However, these shifts in technologies have often taken place gradually and incrementally.

In this section, we organize the innovations in the energy sector that are likely to affect the demand for labor and workforce skills into four areas:

1. increasing productivity in energy extraction
2. reducing environmental damage and emissions of carbon dioxide ($CO_2$)
3. integrating renewable energy into the electricity grid
4. improving end-use efficiency.

**Increasing Productivity in Extracting Fossil Fuels and Nuclear Energy**
According to interviews with academic researchers, for the near term (through 2020), coal and natural-gas extraction will remain major activities in SWPA. Coal-fired power plants and hence demand for coal will remain an important component of Pennsylvania's energy industry as long as the price of coal-fired electricity from SWPA's existing power plants remains lower than new natural gas–based electricity. In the longer term, generators are likely to use substantially more natural gas as coal-fired generating capacity is retired and because policies to reduce emissions of greenhouse gases (GHGs) discourage construction of new coal-fired power-generating plants.

Nuclear energy continues to be an important source of electricity for the region, but utilities are unlikely to construct new nuclear-power plants in SWPA in the foreseeable future. New nuclear energy is currently more expensive than coal or natural gas. Although new nuclear energy does not suffer from intermittency, financing these large, capital-intensive projects adds substantially to the delivered costs, and public concerns about safety are likely to seriously limit construction of new nuclear facilities. Pennsylvania utilities are unlikely to build new nuclear-power plants, but SWPA is a center of the nuclear-power plant manufacturing industry. One of the largest companies in the region, Westinghouse Electric, provides fuel, services, technology, plant design, and equipment for commercial nuclear electric power worldwide. SWPA is also an important center for manufacturing equipment for oil and gas extraction and transportation. The metallurgical industry produces pipes and other products used by this industry.

In contrast, to increase use (and output) of natural gas, the share of electricity from renewable energy will likely remain small in SWPA. Currently, coal and natural gas generate electricity more cheaply than renewable energy; furthermore, solar and wind energy suffer from intermittency problems. Experts see little growth in output from firms specializing in these sources of energy.

Within our focused review, we found that, across fossil and nuclear energy sources, improvements in efficiency and productivity have been a key driver of technological innovation. Table 2.1 summarizes which technologies or novel innovations are propelling change in the productivity and efficiency of energy extraction or production from these sources.

### Reducing Environmental Damage and Emissions of Carbon Dioxide

Increased concerns about environmental impacts brought about by energy production and consumption (including emissions of $CO_2$) have spurred efforts to develop technologies to reduce environmental damage associated with the production chain for fossil resources, operating risk and long-term waste management for nuclear energy, and land use. Because of uncertainty regarding the form and extent of policies to reduce emissions of $CO_2$, the U.S. government is funding most of the research and development supporting this driver of technological change.

**Table 2.1**
**Innovative Technologies in the Energy Sector**

| Energy Type | Technology |
|---|---|
| Fossil | Hydraulic fracturing and directional drilling<br>Advanced geologic surveying<br>Advanced imaging and computer systems for characterizing fossil-energy resources<br>Catalysts facilitating gas separation |
| Nuclear | Small modular reactors |

SOURCES: Arvizu, 2011; Kauffman and Fisher, 2012; Center for Energy Workforce Development, 2012; World Coal Association, undated; Nuclear Energy Institute, 2012; Bartis et al., 2005.

These are some examples of innovations that this driver has enabled (Arvizu, 2011; Kauffman and Fisher, 2012; Center for Energy Workforce Development, 2012; World Coal Association, undated; Nuclear Energy Institute, 2012; Ortiz, Samaras, and Molina-Perez, 2013):

- advanced fossil-fueled power plants that increase efficiency and reduce emissions
- technologies to capture emissions of $CO_2$ from both coal-fired and natural gas–fired power plants and the advanced technologies supporting them
  - carbon capture and storage (CCS) technologies, such as post-combustion capture from flue gas using an amine solvent and chilled ammonia and precombustion capture using integrated gasification combined cycle (IGCC) to isolate and capture $CO_2$ before it is released
  - coal liquefaction and gasification
- oxygen–coal combustion to increase the concentration of $CO_2$ in the flue-gas stream
- oxygen-transport membranes, developed for use in coal-based advanced gasification configurations and capable of efficiently separating oxygen from air.

### Integrating Renewable Energy into the Electric-Power Grid

Renewable energy resources have the potential to reduce significantly the GHG and criteria-pollutant emissions of the utility industry but have their own challenges. Solar and wind systems in particular are intermittent and can create challenges with balancing generation and load, especially as renewable penetration increases. The public and private sectors are investing significant resources in developing and deploying advanced automatic control systems that can detect instabilities and quickly react to maintain power quality, energy-storage systems, and other key technologies. Such technologies include grid-sensing and market-pricing systems to ensure that intermittent resources are priced accordingly (Anadon et al., 2011). Some technological innovations in these energy industries are listed in Table 2.2.

Table 2.2
Innovative Technologies for Integrating Renewable Energy into the Grid

| Category | Technology |
|---|---|
| Advanced automatic control systems | • Sensors and controls that sense instabilities and react quickly to maintain power quality<br>• Power electronics that enable control of the electricity grid and that facilitate the integration of renewable resources |
| Market-pricing mechanisms | • Advanced modeling to assess operational risks of intermittent power and the value of mitigating the potential for outages |
| Wind | • Wind-turbine system and components<br>• Wind resource modeling and forecasting<br>• Power electronics to integrate wind resources into the grid<br>• Offshore wind farms and advanced small-scale wind turbines<br>• Advanced gearboxes |
| Solar | • Coatings, glass, and balance of systems<br>• Improved PV technology: thin-film cells and modules, nano-material-enabled technologies, and advanced manufacturing techniques<br>• Concentrating solar power: low-cost, high-performance thermal storage; advanced absorbers, reflectors, and heat-transfer fluids |

SOURCES: Arvizu, 2011; Kauffman and Fisher, 2012; Center for Energy Workforce Development, 2012; World Coal Association, undated; Nuclear Energy Institute, 2012; Ortiz, Samaras, and Molina-Perez, 2013.

NOTE: PV = photovoltaic.

## Improving Efficiency of Energy Use

Another driver for technological innovation is the need to use energy more efficiently. Smart building designs and systems, small-scale energy-storage systems, and electric vehicles are all technologies that fall into this category. To be able to install and maintain such systems may require expertise across many domains, including electronics and power electronics; industrial controls; information technology; construction; and heating, ventilation, and air conditioning. Examples of innovations designed to enhance end-use energy efficiency are shown in Table 2.3.

Table 2.3
Innovative Technologies for Improving Efficiency of Energy Use

| Category | Technology |
|---|---|
| Smart building design | • Advanced active systems and passive-design concepts (e.g., computerized building energy optimization and simulation tools, electrochromic windows, building-integrated PV products and PV-thermal arrays, compressorless refrigeration cycles)<br>• Fully integrated building systems<br>• Smart systems that can sense occupant behavior and needs, configure themselves automatically, and report problems automatically |
| Energy-storage systems | • Advanced batteries and management systems<br>• Thermal storage systems |
| Advanced vehicles | • Fuel utilization (e.g., advanced fuel chemistry and testing, engine–fuel interactions)<br>• Advanced power electronics<br>• Vehicle systems (e.g., advanced heating and cooling, vehicle thermal management)<br>• Home charging systems and the vehicle-to-grid interface<br>• Lightweight vehicle materials<br>• Vehicle automation |

SOURCES: Arvizu, 2011; Kauffman and Fisher, 2012; Center for Energy Workforce Development, 2012; World Coal Association, undated; Nuclear Energy Institute, 2012.

## Skills Needed to Keep Pace with Technological Innovation

### Workers Need Specific Content Knowledge

Experts in energy policy, technology, and innovation whom we interviewed described key developing technologies and the accompanying skill sets that may be needed. Because many of the innovations in the energy sector are aimed at more efficiently extracting natural gas or coal or reducing environmental impacts, several experts noted the importance of exposure to geology and environmental science for semiskilled workers and junior engineers to provide contextual relevance for their work in the industry. Courses on oil- and gas-extraction processes and methodologies, as well as geology and hydrology, were recommended for semiskilled workers, such as technical or engineering assistants, to provide a comprehensive understanding of this line of work. One researcher suggested that receiving training on both intelligent building designs and technologies and information technology would be beneficial because these domains are becoming increasingly integrated.

Several innovations apply sensors and controls to improve system performance. As a result, several of our interviewees noted that semiskilled workers are increasingly expected to be able to read sensors and interpret data to identify potential problems in machine performance and inefficiencies throughout the system. In response to the growth of natural-gas production, researchers anticipate development in technologies for environmental monitoring of water during hydraulic fracturing processes. Semiskilled workers in SWPA need to learn how to operate new instruments, including sensors used to detect pollutants in the water, and to take action to mitigate potential damage in real time.

Demand for workers who can operate and manage these technologies may be greater than for those with the very specific technical skills needed to manufacture and repair these products. Several experts claimed that these technologies do not necessarily translate into a need for a radical shift in the skill sets of the workforce. Even if there is a rise in wind power, for instance, the new practices do not dramatically change the qualifications needed from semiskilled workers who operate and maintain wind farms, but an increase in wind generation may require workers to read and interpret more sensors and instrumentation. A solar-energy expert explained that working with new grid architecture would likely require an electrician to learn new codes and regulations but would not necessarily affect the skill set needed. Electricians could learn such codes and regulations themselves by reviewing training manuals or documents about the new codes or by enrolling in a short community-college course. Growth of PV installations may require current electricians to expand their skills and knowledge of codes and standards.

Interviewees admitted that it is difficult to gauge how different the technology used in the electric-power industry will be in 2020. Although advanced manufacturing techniques will be instrumental in the continued development of PV systems, manufacturing such systems is highly automated, so the implications for the skills and training of the workforce are unclear.

When asked to provide a few examples of skills that may be needed one decade from now, one interviewee identified understanding industrial processes as a key competency in the natural-gas industry. Liquefaction and gas-processing systems involve multistep freezing and separation processes that are highly integrated and sensitive to many parameters. The result is the need for a workforce with backgrounds in mechanics, pipe-fitting, information technology, maintenance, and process chemistry.

Other technology areas widely cited by experts as providing the potential for significant changes in the skills demanded of the work-force included mechatronics, integrated system technology, additive manufacturing, machining and fabrication, welding, metrology, and power electronics.

## To Keep Pace with Technology, Workers Need Hands-On, Transferable Skills

All directors of regional training and education programs and business and nonprofit leaders with whom we spoke for this study emphasized the need for workers to be agile. Several discussions revealed the common notion that, overall, technological innovations in the energy industry are yielding incremental changes in the skill-set demands on the semiskilled workforce. Though the energy industry is known to be constantly evolving and advances are becoming increasingly technical, educators believe that the workforce faces issues that have more to do with refining existing core skills than with developing entirely new ones. Interviewees agree that certain standard competencies and certifications are required in the industry before employees pursue any type of specialized training related to the energy industry.

Construction-industry apprentices, for instance, are required to demonstrate competency in six hands-on disciplines to prepare them for the workforce:

- welding
- cutting steel with an oxyacetylene torch
- tie-reinforcing steel
- climbing a steel column

- using surveying equipment with precision
- rigging.

Manufacturing and automation, on the other hand, place special emphasis on skills in digital technologies, writing sequences of operations, and energy management. These skills are valued for their portability and adaptability in the industry.

### Workers Need Technical and Behavioral Competencies

Employers with whom we spoke described desirable competencies they seek when hiring new candidates. They also discussed the kinds of on-the-job training they provide to ensure that employees are well equipped for these evolving roles.

The natural-gas industry requires a broad range of skills. One industry leader noted needs for mechanical, chemical, and civil engineers; electricians; mechanical technicians; environmental technicians; chemists; biologists; and hydrologists. When discussing the recruitment process to fill such positions, one interviewee stated that a candidate who holds a four-year engineering degree relays a signal to the employer that he or she has a fundamental level of understanding of engineering principles. Candidates with computer technology backgrounds are also highly regarded, given the prevalence of such systems.

In addition to ensuring that a candidate meets core technical competencies, a firm assesses each candidate's behavioral competencies during the recruitment and interview process to ensure that he or she has the capacity to undergo the extensive training needed to integrate him or her into the company's operations. Several industry professionals emphasized the importance of recruiting employees who are flexible and adaptable and have a strong work ethic in order to keep up with constantly changing job demands. Other professional traits, such as consistency, integrity, the ability to follow directions, and leadership, were also highly sought. Once these core requirements are met, one employer with whose representative we spoke provides extensive hands-on professional and technical training and development. A recently hired electrician in this company, for instance, will undergo a week of training to learn the electrical components of power stations. Instructors contracted from major universities, as well as local training centers and community colleges, are invited to provide training on various topics (e.g., new gas-separation technologies in the gas industry). As for professional development, training sessions on topics, such as leadership, management, and conflict resolution, are also offered for any salaried personnel at different points in their careers.

# CHAPTER THREE

# How Education and Training Can Successfully Adapt to Changing Labor-Market Needs

In this chapter, we summarize the findings from our review of U.S. and international education training systems. We identified exemplary systems through a review of Organisation for Economic Co-operation and Development (OECD) publications that have documented the quality of training programs. There is extensive literature on the quality of vocational programs in OECD countries because these countries tend to have well-developed TVET systems. Informed by our review, we selected highly regarded programs in the United States and other countries for closer study. The programs we selected came from Texas, South Carolina, Germany, and Finland. The four exemplary programs had documented success in adapting to labor-market changes and in graduating students who have found employment in their field of study. We also looked at individual community colleges in the United States that have successfully set up specific certification or training programs in response to labor-market needs in specialized industries, such as health care and the automotive industry. In our search, we attempted to locate evaluations of education and training program efforts that have successfully aligned programs to the energy sector's workforce skill demands. However, few papers addressed the energy industry specifically, and none of them was a rigorous evaluation study. Thus our review is based on the broader literature.

From this review, we synthesized five promising practices that characterize education and training programs that have reportedly been successful in meeting employer workforce and skill demands, particularly when confronted with the challenge of evolving employer needs:

- Anticipate occupational demands.
- Adopt quality curricula.
- Implement a blended instructional approach that includes work-place learning.
- Hire and retain quality instructors.
- Engage in meaningful, continuous, quality-assurance processes.

## Anticipate Occupational Demand

Our review revealed that successful education and training programs increase their responsiveness to labor-market needs by establishing mechanisms to anticipate occupational demands and skills, including any changes resulting from technological innovations.

The literature revealed two mechanisms that education and training programs should incorporate (Elkins, Krzeminski, and Nink, 2012; OECD, 2011a, 2011b):

### Systematically Involve Employers and Regional Partners in Setting Policies, Training Content, and Qualifications

Employers and regional partners can be involved in a variety of ways, largely depending on the extent to which the federal, state, and local governments are involved in vocational, or career and technical, education. In countries where the national government oversees TVET systems, stakeholders are involved through participating at national-level councils to plan and develop requirements. Other countries, such as Germany, where responsibility for vocational training is divided between the federal government and localities have developed close partnerships among government, employers, and trade unions at all levels. For example, regional nongovernmental-organization partners in Germany participate in national boards and state and regional committees, on which they are able to influence the curriculum of vocational and technical education and ensure that their requirements and interests are taken into account. In decentralized educational systems, involving nongovernmental organizations ends up being the responsibility of each local training institution.

Individual education and training institutions that focus on vocational and technical careers and that have demonstrated an ability to develop and nurture sustained relationships with employers are those that do the following:

1.  Assess market needs and future trends.
2.  Identify means for responding to those needs.
3.  Identify appropriate partners.
4.  Find new avenues to market their services to employers and convey messages to potential partners about their quality as training providers, their flexibility, and their problem-solving approaches.
5.  Develop relationships with an expanded set of constituencies, including a broader range of employer partners and economic-development groups, community-based organizations, K–12 systems, four-year educational institutions, political leaders, industry associations, and unions (MacAllum, Yoder, and Poliakoff, 2004).

### Engage in Ongoing Research Activities and Survey Employers to Assess Demand for Skills and Quality of Graduates

Successful postsecondary TVET systems use well-designed labor-market studies to identify demand for skills and assess the quality of their graduates (OECD, 2010). Such systems within the United States utilize job projections from the U.S. Bureau of Labor Statistics, survey their graduates and employers, and meet with industry associations and employers to discuss their needs. Training systems need to be in touch with local job markets, as well as follow regional trends, to ensure that programs are relevant to the needs of the regional labor market, as well as to their graduates' needs to be employable and find jobs.

## Adopt Quality Curricula

Successful TVET systems put in place curricula that develop workers' broader skills that are transferable across positions and respond to changes in job responsibilities that result from technological innovation, as well as occupation-specific skills (OECD, 2010; Elkins, Krzeminski, and Nink, 2012). Having both transferable, broad skills and occupation-specific skills enhances workers' abilities to remain employed and develop their careers. For example, many current jobs across various sectors, including the energy sector, use computers or computerized instrumentation that require those operating these instruments to be able to read the data generated and make the appropriate decisions, as well as communicate data results and reports to their superiors. Such technological innovations have made problem-solving, critical thinking, and complex communication skills critical. The development of such skills requires solid preparation in language skills and mathematical literacy (Kézdi, 2006; Levy and Murnane, 2004). In sectors facing rapid technological change, the ability to learn is crucial, and employers highly value the generic skills underlying this ability (Smits, 2007; Köllő, 2006).

Vocational training curricula should also address the development of soft skills, such as the ability to work in teams, communication skills, entrepreneurship, dealing with customers, and work discipline (OECD, 2010). Studies have shown that professionalism and maturity are among the least developed skills among new employees entering the workforce directly from high school, followed by basic reading, writing, and social skills (Kézdi, 2006).

## Implement a Blended Instructional Approach That Includes Workplace Learning

Our review revealed that balancing in-class educational training with workplace (e.g., on-the-job) training is noted as highly important. In-class education develops general skills, such as literacy and numeracy, occupation-specific content knowledge, and theoretical knowledge about the career (OECD, 2010). Workplace learning, on the other hand, promotes the development of technical performance–based skills on modern equipment and supports the development of soft skills through real work experience (Aarkrog, 2005; Lasonen, 2005). Because of ongoing innovations in technology, the acquisition of hard skills often requires practical training on expensive equipment that only companies operating in the sector can provide. Technologies often change rapidly, and equipment quickly becomes obsolete. Therefore, community colleges or other educational institutions are often unable to afford modern equipment or to continually replace outdated equipment (OECD, 2010). Where equipment is expensive or dangerous, simulated work environments in a centralized educational setting (for example, in a center of excellence) may be more cost-effective than purchasing and operating an expensive machine (OECD, 2010).

## Hire and Retain Quality Instructors

The quality of teaching and training professionals is critical to effective learning in TVET programs. Instructors must have both pedagogical skills and work experience. Continued education and training need to be provided to help instructors continually improve their pedagogical skills and upgrade their skills to keep up with changes in occupations and the technologies used.

Some exemplary education and training programs included in our review select instructors who also work in the industry (OECD, 2010). These programs tend to offer competitive salaries to attract industry professionals into the teaching profession. The advantage of this model is that instructors remain informed about changes in occupational knowledge and skills because of their connections to their industries. Other programs we reviewed hire instructors who have obtained the appropriate industry certification or who have passed a state- or federally approved competency test in the specific occupation or sector. However, passing a certification examination does not guarantee high-quality teaching. Professional-development programs need to train instructors in teaching techniques and in finding ways to make their teaching applicable to the workplace (OECD, 2010).

## Engage in Meaningful, Continuous Quality-Assurance Processes

Countries use different strategies to ensure the quality of their TVET programs. Defining graduation or qualification requirements for each occupation at the national level or through regulated occupations and professional or industry associations is one strategy. Another strategy is to insist that each program engage in an accreditation process whereby the institution is evaluated both externally and internally on defined measures throughout the process (OECD, 2010). Many quality-assurance processes require institutions to engage in ongoing internal review to monitor their performance before they are externally evaluated again. A third strategy is to have training institutions maintain websites with detailed information regarding their programs and performance, thus enabling market pressures to provide quality assurance.

# CHAPTER FOUR

# Case Studies: Education and Training Providers in Southwestern Pennsylvania

We used the promising practices detailed in Chapter Three to examine four education and training programs in SWPA that prepare students for semiskilled jobs in the energy sector.

The purpose of this case-study analysis is to determine whether and to what degree current training systems incorporate some of the promising features noted in Chapter Three, identify areas in need of improvement, and understand the challenges that training systems in the energy sector face. We purposefully selected the institutions to be representative of the wide range of types of energy-specific education and training providers and for their geographic locations across the 32-county SWPA region. The case-study sites included one community college, one institution of higher education that offers both bachelor's and associate's degrees, one union apprenticeship program, and one postsecondary career- and technical-education center. Descriptions of these programs are provided in Table A.1 in the appendix. Further details on the case-study site methodology we employed and site selection are available in Chapter One.

This chapter is organized by our findings on whether case-study sites incorporated the promising practices identified in Chapter Three as key to supporting labor-market needs in an evolving or fast-changing labor market: anticipate occupational demand; adopt quality curricula; implement a blended instructional approach; hire and retain quality instructors; and engage in meaningful, continuous quality-assurance processes.

## Most Sites Need to Formalize and Structure Strategies to Anticipate Occupational Demand

Of the four education and training programs included in the study, most have established some type of advisory committee, group, or council to help anticipate changes in occupational demands and skill sets in the energy sector. The membership of advisory committees often includes representatives from industry consortia or specific firms in which the program aims for its graduates to find employment.

These advisory committees work in a variety of ways. In one training program, a college department provided technical training and support to high-school teachers who instructed their students in technologies involving the natural-gas industry. Any changes that occurred within the college course curriculum were automatically transferred to the high-school level. To anticipate changes in demand for certain skill sets, the advisory committee for this program, made up of faculty from various disciplines associated with the natural-gas industry, met every three years to revise the curriculum. Other training programs and a community college that provides advanced technical training on manufacturing use their advisory committees to formally design the curriculum of the program through a design and curriculum (DAYCUM) process. As part of this process, committee members, who include business and community partners, meet to modify components in an existing training curriculum to match the current needs of the industry. Though these committees play an instrumental role in curriculum development, it is unclear how often they meet, when exactly they meet, and the exact process by which curricula are modified. Another program director currently working to launch a new training facility and develop an entire curriculum also employed the DAYCUM process.

Although the literature we summarize in Chapter Three identifies mechanisms that encourage industry–educator partnerships as an effective way to detect changes in market trends and demands, we found that the training programs under study did not systematically engage employers. Among our sites, the process for industry–educator partnerships tends to be employer-initiated: Employers would initiate contact with the program directors to inform them about specific workplace skills that need to be developed for potential employees. These interactions with employers seem to operate in an informal manner through quick phone calls rather than through a formalized, regular dialogue. Nonetheless, program directors reportedly use these exchanges of information to adapt training processes or content as a way to enhance current education and training programs. Only one program among our case studies stated that such conversations are documented and tracked so they can be preserved and the recommendations implemented in a timely fashion.

The literature also identifies using data on labor-market needs as a strategy to assess demand. One training program has an active research organization that examines student employment placement, which allows it to evaluate the current program and modify it accordingly. None of the other case-study programs seems to be engaging in ongoing research to understand local labor-market needs.

## Sites Integrate General Skills and Content Knowledge into Their Curricula but Need to Integrate Workplace Competencies

Most program directors with whom we spoke emphasized the importance of developing soft skills. They noted that their business partners have expressed appreciation to these training programs for their instruction on general skills to their students, such as writing resumes, interviewing, leadership, taking criticism, expectation-setting, working in groups, overall professional conduct, and safety, which are highly desirable and necessary in the industry. Training and education directors with whom we spoke unanimously supported the continued development of soft skills among semiskilled workers. As a result, the training programs have incorporated an array of soft skills, ranging from resume writing, interviewing, work ethics and integrity, professionalism, communication and cooperation, and leadership, into their program curricula. Though these are skills that training programs currently teach their students, program directors indicated that these skills would remain relevant no matter the future directions of the industry.

Programs varied in terms of how they delivered instruction in soft skills. Although some programs set aside time for explicit instruction on professionalism, other programs integrated these lessons throughout the class. Only one training director felt that it was too difficult to integrate these behavioral competencies explicitly into the curriculum, though she agreed that they are important, especially when it comes to teamwork and critical thinking. With respect to technical skills, program directors emphasized the importance of students acquiring a "core knowledge and foundation" of the industry, as well as a skill set that was portable. They identified math and science, facility with digital technologies, the ability to write sequences of operations, and energy management as core competencies in the energy sector that could be transferred between jobs in manufacturing and automation. Other key skills, especially in oil and gas development, included welding, rigging, and operating surveying equipment.

## Sites Incorporate Blended Learning, but Opportunities for Workplace Experience for Students Are Missing

All of the sites we visited incorporated a blended approach to learning, including both in-class and laboratory settings in which students learn how to operate machinery. When we visited training centers, program directors provided tours to show the types of machinery, equipment, and tools that students use when rigging, welding, post-tensioning, doing simulated modeling exercises, and fabricating during hands-on learning. Mechatronics, which combines electrical and mechanical skills into one specialty, was repeatedly cited as an emerging area of focus across training sites. One training program, which is taught in high-school laboratories, consisted of about 20 to 30 percent theoretical content, and the rest consisted of hands-on application, while others were said to be a 50/50 balance of theory (much of which pertains to safety) and application.

However, none of the training programs currently provides students with workplace experience, such as internships. Some sites are in the midst of creating community partnerships to establish internship and externship programs, while others are constructing new training facilities to be regionally and strategically located.

Two program directors want to adopt alternative media formats (such as online videos or audio recordings) so students can learn theoretical aspects of the energy industry on their own schedule and have more time in the facilities for in-depth, hands-on training in the classroom. Some of these facilities host blended programs. For example, one school has one of the greenest kitchens in the world, in which culinary students are being trained in preparing food.

## Hiring and Retaining Instructors Continues to Be a Challenge

All interviewees agreed that employing instructors with direct experience and expertise in the industry would yield the best learning experience for students. In addition to communicating expectations in the workplace to students, these instructors are also better able to teach core curriculum courses that are transferable across disciplines. The programs are especially eager to seek out professionals with specialized skill sets and expertise in their trades.

Programs use professional and academic networks, former retirees, and general hiring strategies to identify potential candidates for teaching. Although some retirees have been hired as full-time faculty at some training programs, other education centers hire industry experts for one-day courses to teach students specific skills. Recruiting nonretiree professionals to teach is difficult for training programs because the monetary compensation for teaching often cannot match that offered by industry. Using short-term instructors provides students with current information, as well as an opportunity to build their professional networks.

Several education and training programs provide support to foster professional development among instructors. Some provide instructors with workplace experience with training in teaching. Program directors said they assisted teachers with developing syllabi and properly evaluating their classes with Classroom Assessment Techniques (CATs).

Programs conduct evaluations to ensure quality standards of instruction, assign fellow faculty members to serve as mentors, and host staff meetings. Not all of these strategies are pursued at every site; only one program explicitly mentioned providing teachers with continuing-education opportunities and updates for instruction and content within their fields. One program stated that instructors attend several training courses throughout the year, along with one mandatory week of schooling that focuses on how to be a professional educator. This course instructs teachers in soft skills, teaching techniques, and improving classroom efficiency.

One program reached out to high-school teachers providing information on gas-industry content knowledge that high-school teachers could teach their 10th-, 11th-, and 12th-grade students about the energy industry.

## Quality-Assurance Processes Are Considered Important

All directors had quality-assurance strategies involving accreditation, curriculum development, safety, and internal program operations. Three program directors stated that, to earn accreditation, facilities must have continuing-improvement strategies in place. One program director noted that he works with an auditing team called the International Association for Continuing Education and Training, which provides a list of requirements to be fulfilled in order to gain accreditation as a certified training facility; another director worked with the National Alliance of Concurrent Enrollment Partnerships. Another center collected data on six different measures (such as key performance indicators and placements) and then analyzed them statistically to create a system that would engender improvement. The director asserted that such a rigorous tracking methodology was necessary in industries in which technologies constantly evolve.

In terms of ensuring the quality of educational curriculum, one program's director plans to update the use of DAYCUM with process and task analysis as a model to develop and update the curriculum. DAYCUM and process and task analysis use a group of stakeholders from energy industries to provide input on skills, knowledge, and abilities that should be taught in the program. Advisory committees and other stakeholders were mentioned by representatives from three sites as a strategy to gather experts and receive feedback on curriculum development, as well as workforce needs in the energy sector. One program director sought assistance in identifying areas for improvement, especially from a teaching perspective, in a high-quality engineering technology program. Individuals from two sites said that modifications to the curriculum were made internally; instructors work together to identify patterns in student examinations to gauge whether specific subjects needed more attention than others and adjust instruction accordingly. One program used college faculty members to write and grade final exams as a way to provide a higher-level assessment of student learning. Two training centers use student evaluations to elicit feedback on general program operations.

Only one program director explicitly stated that federal entities, such as the U.S. Department of Labor and the Occupational Safety and Health Administration, conduct periodic audits to monitor labor regulations and ensure safety standards.

CHAPTER FIVE

# Ensuring That Education and Training Meet Evolving Energy-Sector Needs in Southwestern Pennsylvania

Based on our analysis of the literature on promising practices to support education and training programs' development of talent and our case studies of Pittsburgh-region programs' characteristics, this chapter outlines five suggested measures to promote education and training providers' abilities to successfully train talent in the skills needed in the energy sector in the SWPA region and to accommodate innovation in the sector. The measures listed here provide general guidance to key education and training providers, business leaders, and nonprofit stakeholders but stop short of providing explicit, detailed implementation plans. Instead, we intend these measures to be a springboard for further discussion among key stakeholders in the region about the feasibility of enacting the measures, details on how to implement the measures, and the roles and responsibilities for both industry leaders and education and training providers in collaborative efforts.

## Develop Sustained Partnerships Between Industry Leaders and Training Providers

Industry partners and employers can play a critical role in supporting education and training providers through involvement with curriculum development, identifying job needs and vacancies, advertising programs, providing feedback, and providing equipment and materials for training. The key to success is *sustained* and *continuous* involvement by industry leaders.

In this section are some ways in which regional industry leaders can engage with education and training providers:

- Forecast demand for occupations. Provide regular updates on jobs that need to be filled and position requirements.
- Provide regular input on curricula in terms of content knowledge, hands-on skills to promote the flexibility of workers, workplace competencies, and soft skills.
- Support the acquisition of equipment and supplies to make the classroom environment as close to the on-the-job experience as possible. This can be done by establishing strategically located laboratory centers or centers of excellence in which several training providers share space and equipment provided by industry employers.
- Integrate workplace learning. Embed formal internships and apprenticeship programs within training programs so workers are familiar with workplace behavioral requirements and the on-the-job skills needed.

## Integrate Technical, Occupation-Specific Training with Workplace Readiness and Other Soft Skills

Incorporating soft skills in a training program's curriculum helps to prepare graduates for on-the-job demands. Integrating technical training that conveys specific content knowledge along with work readiness and basic skill training (coupled with individual case-management services) allows students in the training programs to obtain the wide range of content and skills necessary to be competent, yet agile, employees who are in demand. Including formalized internships or apprenticeships with partner employers supports the development of these skills.

## Encourage the Recruitment and Retention of Quality Instructors

Instructor retention can be challenging because instructors with experience are often lost to private-sector employers offering higher wages. To ensure that education and training providers can hire instructors who have both workplace and teaching experience, we suggest two strategies:

- Institute continuous professional development for current instructors, which could include time in the workplace environment of partner employers in the energy sector.
- Industry leaders should collaborate with training providers to provide incentives to their employees to become instructors, such as these:
  - Offer a rotating visiting-instructor position to their employees.
  - Offer time off for employees who choose to teach with an open-door policy for these employees to return to work full time.
  - Suggest or refer recent retirees to education and training providers as possible instructors.

## Incorporate Continuous Quality Assurance into Programs

Accreditation is the first step to ensure quality; all education and training programs need to be accredited. However, accreditation checks to ensure that an institution complies with general standards often occur only every few years. Given the pace and scope of innovation and changing labor-market needs in the energy sector, education and training providers need ongoing input to monitor whether their programs are meeting the energy industry's needs. Consequently, accreditation needs to be complemented with other strategies to provide continuous feedback on ongoing performance, successes to date, and information on what is working. If programs are *not* graduating employable talent, they need to quickly determine why students are not graduating. This can take the form of an oversight committee from an external body, which could include industry, government, nonprofit, and other community stakeholders. Programs can also incorporate course assessments and ongoing monitoring.

A continuous feedback mechanism has a variety of benefits:

- It allows the organization to ensure that organizational practices are current and are developed in line with the most-recent scientific research.
- The education or training provider can monitor progress toward goals. This can be done by combining objective criteria with survey data of students and employers for more-comprehensive measurement or by evaluating policies, procedures, and related efforts to determine impact on goals.
- Action can be taken quickly in response to student or employer concerns.

A mechanism to incorporate continuous quality assurance provides an opportunity for the education and training provider to determine whether processes are working, programs and initiatives are reaching intended outcomes, and goals (such as employment of graduates upon graduation or after a specified length of time) are being met.

## Document Progress on Whether Regional Training and Employment Goals Are Being Met

We suggest that education and training providers and industry leaders in SWPA collaborate to develop annual reports summarizing progress toward dual goals in program development: the intake of students into programs and hiring and retention rates of graduates from programs by regional industry. Annual reports would provide lessons learned if goals have not been met and collaborative plans for the upcoming year. These types of reports can be made available to the public to encourage further feedback regarding how education and training efforts, industry collaborations, and other processes can be improved so the region can best meet employment goals.

When conducting an evaluation, one can examine two areas:

- *implementation*, which determines whether programs are being implemented as designed and whether inputs are sufficient to ensure that the program can be provided to the intended populations (e.g., students, graduates, and employees)
- *outcomes*, which measure the extent to which the programs and collaborations are meeting goals, are effective, or are making a difference.

There is a clear need for information in the region on whether education and training programs and other initiatives are having an impact. Having these data on hand would allow the region's stakeholders to track the effectiveness of ongoing relationships and training programs in meeting employment, hiring, or retention goals for the region. Evaluating and tracking ongoing progress can help inform decisions about where to focus resources or prioritize future investments in training programs and determine return on fiscal and personal time investments in training.

## Concluding Remarks

In response to economic and environmental challenges, the U.S. energy sector has become highly innovative in developing and applying new technologies and operating approaches. Because innovation may require more–highly skilled labor to manufacture and deploy energy technologies and systems, the energy workforce will need to improve its skills as demand for labor shifts. These developments highlight the need for training and educational institutions to be well equipped to anticipate technological advances and their implications and to facilitate the retraining of the workforce in new competencies.

Institutions that train, and retrain, workers—among which community colleges and career and technology centers are the most important—face many challenges if they are to assist the workforce to master required new skills and competencies. These institutions need to have the ability to anticipate changes in needed workforce skills as a result of technological innovation in the energy sector and to adapt quickly to changes in demands for workforce skills by modifying their existing curricula and programs and by developing or introducing new ones. These institutions must also develop workers' core skills, as well as directly related nontechnical skills, such as critical thinking and problem-solving. Workers also need to master appropriate information technology to facilitate on-the-job acquisition of new competencies demanded by technological innovations. These nontechnical skills provide workers with the ability to cope and respond to unexpected changes in their work due to innovations and to take advantage of these changes as opportunities to transform their skills.

The suggestions presented in this report are intended to jumpstart discussions among regional stakeholders to ensure that the workforce is trained and educated in ways that best meet the changing labor-market needs of the energy sector.

# Descriptions of Education and Training Facilities Used for Case Studies

Table A.1 lists the case-study sites and their locations, areas of specialty in the energy sector, program sizes, admission requirements, program durations, and program categories.

**Table A.1**
**Descriptions of Training and Education Programs Included as Case Studies**

| Site | Location | Area of Specialization Within Energy Sector | Program Size | Admission Requirements | Program Duration | Program Category |
|------|----------|---------------------------------------------|--------------|------------------------|------------------|------------------|
| A | Outside Pittsburgh | Natural-gas industry: electrical utility technology, occupational health and safety, mechatronics system, petroleum and industrial process operation technology, industrial electricity technology | 6,600 full- and part-time students | High-school diploma or equivalent or demonstration that would benefit from courses | Stackable credit system | Community college: internship component and blended approach |
| B | Outside Pittsburgh | Construction industry: structural steel erection, precast concrete, bridge erection and repair, ornamental iron and curtain wall, concrete and reinforcing steel, wind turbines, welding, rigging and machinery moving | More than 1,800 members; 50 students per class | Minimum age of 18, high-school diploma or equivalent, good physical health, drug and alcohol screening, valid driver's license, access to car, committee review | 3-year program; 280 hours of training | Workforce union: blended approach |

**Table A.1—Continued**

| Site | Location | Area of Specialization Within Energy Sector | Program Size | Admission Requirements | Program Duration | Program Category |
|---|---|---|---|---|---|---|
| C | Downtown Pittsburgh | Carbon management: nuclear-reactor design, wind-turbine manufacturing, materials and sensors, electric-power distribution, shale-gas production, advanced building systems | About 100 students (projected to grow rapidly) | Information not yet publicly available | Information not yet publicly available | Nonprofit training center: blended approach |
| D | 200 miles from Pittsburgh | Natural-gas industry: plastics industry, welding, machinist general and tool technology, electronics and engineering technology, diesel, building construction, collision repair | More than 700 students | Grades 10–12, some prerequisites, testing requirements, and reading-level proficiencies apply | High-school academic year: course enrollment | High-school and college dual and concurrent enrollment: classroom learning |

SOURCES: Program websites and interviews with program directors.

# References

Aarkrog, Vibe, "Learning in the Workplace and the Significance of School-Based Education: A Study of Learning in a Danish Vocational and Training Programme," *International Journal of Lifelong Education*, Vol. 24, No. 2, March–April 2005, pp. 137–147.

Anadon, Laura Diaz, Matthew Bunn, Gabe Chan, Melissa Chan, Charles Jones, Ruud Kempener, Audrey Lee, Nathaniel Logar, and Venkatesh Narayanamurti, *Transforming U.S. Energy Innovation*, Cambridge, Mass.: Harvard Kennedy School, Belfer Center for Science and International Affairs, 2011. As of October 8, 2014:
http://belfercenter.ksg.harvard.edu/publication/21528/
transforming_us_energy_innovation.html

Arvizu, Dan E., *Innovation in the Energy Sector: The Role of Energy Efficiency and Renewable Energy*, National Renewable Energy Laboratory, January 5, 2011. As of May 9, 2013:
http://www.nrel.gov/director/pdfs/20110105_nsf.pdf

Bartis, James T., Tom LaTourrette, Lloyd Dixon, D. J. Peterson, and Gary Cecchine, *Oil Shale Development in the United States: Prospects and Policy Issues*, Santa Monica, Calif.: RAND Corporation, MG-414-NETL, 2005. As of October 14, 2014:
http://www.rand.org/pubs/monographs/MG414.html

Center for Energy Workforce Development, "Gaps in the Energy Workforce Pipeline 2011 CEWD Survey Results," Washington, D.C., c. 2012. As of October 14, 2014:
http://www.cewd.org/surveyreport/CEWD-2011surveyreport-021512.pdf

Committee on Emerging Workforce Trends in the U.S. Energy and Mining Industries; Committee on Earth Resources; and Board on Earth Sciences and Resources Division on Earth and Life Studies, in collaboration with Board on Higher Education and Workforce Policy and Global Affairs Division, *Emerging Workforce Trends in the U.S. Energy and Mining Industries: A Call to Action*, 2013. As of August 31, 2013:
http://www.nap.edu/catalog.php?record_id=18250

Commonwealth Economics, *Energy in Pennsylvania: Past, Present, and Future*, Commonwealth of Pennsylvania Department of Environmental Protection, February 2013. As of November 11, 2013:
http://www.elibrary.dep.state.pa.us/dsweb/Get/Document-96943/Final%20PA%20Comprehensive%20Energy%20Analysis.pdf

Elkins, Jill, Chris Krzeminski, and Carl Nink, *Labor Market Analysis Leads to Demand-Driven TVET Programs*, Centerville, Utah: Management and Training Corporation, 2012.

Kauffman, Jim, and Laura Fisher, *Workforce Analysis Report: Energy Sector Jobs in Greater Pittsburgh—Executive Summary Report*, Pittsburgh, Pa.: Allegheny Conference on Community Development, August 30, 2012. As of October 8, 2014
http://alleghenyconference.org/PDFs/Misc/WorkforceAnalysisReport083012.pdf

Kézdi, Gábor, *Not Only Transition: The Reasons for Declining Returns to Vocational Education*, Central European University and IE/HAS, June 2006. As of October 8, 2014:
https://www.cerge-ei.cz/pdf/gdn/rrc/RRCV_15_paper_01.pdf

Köllő, János, *Workplace Literacy Requirements and Unskilled Employment in East-Central and Western Europe*, Budapest Working Papers on the Labour Market 2006/7, Budapest: Institute of Economics, Hungarian Academy of Sciences, December 2006. As of October 8, 2014:
http://mek.oszk.hu/06300/06321/

Lasonen, Johanna, *Workplace as Learning Environments: Assessments by Young People After Transition from School to Work*, Budapest Working Papers on the Labour Market 7, 2005. As of October 8, 2014:
http://www.bwpat.de/7eu/lasonen_fi_bwpat7.pdf

Levy, Frank, and Richard J. Murnane, "Education and the Changing Job Market," *Educational Leadership*, Vol. 62, No. 2, October 2004, pp. 80–83.

MacAllum, Keith, Karla Yoder, and Anne Rogers Poliakoff, *The 21st-Century Community College: A Strategic Guide to Maximizing Labor Market Responsiveness*, Vol. 1: *Unleashing the Power of the Community College*, Washington, D.C.: U.S. Department of Education Office of Vocational and Adult Education, September 2004.

National Mining Association, "Clean Coal Technology," Washington, D.C., undated. As of May 10, 2013:
http://www.nma.org/pdf/fact_sheets/cct.pdf

Nuclear Energy Institute, "Innovation-Enhances-Safety-and-Reliability-at-Americas-Nuclear-Energy-Facilities," *Nuclear Energy Institute*, Summer 2012. As of May 10, 2013:
http://www.nei.org/News-Media/News/News-Archives/
innovation-enhances-safety-and-reliability-at-amer

OECD—*See* Organisation for Economic Co-operation and Development.

Organisation for Economic Co-operation and Development, *Learning for Jobs: Synthesis Report of the OECD Reviews of Vocational Education and Training*, Paris, August 10, 2010. As of October 8, 2014:
http://dx.doi.org/10.1787/9789264087460-en

———, *OECD Reviews of Vocational Education and Training: A Learning for Jobs Review of the United States, South Carolina 2011*, Paris: OECD Publishing, January 30, 2011a. As of October 8, 2014:
http://dx.doi.org/10.1787/9789264114012-en

———, *OECD Reviews of Vocational Education and Training: A Learning for Jobs Review of the United States, Texas 2011*, Paris: OECD Publishing, February 28, 2011b. As of October 8, 2014:
http://dx.doi.org/10.1787/9789264114029-en

Ortiz, David S., Constantine Samaras, and Edmundo Molina-Perez, *The Industrial Base for Carbon Dioxide Storage: Status and Prospects*, Santa Monica, Calif.: RAND Corporation, TR-1300-NETL, 2013. As of October 8, 2014:
http://www.rand.org/pubs/technical_reports/TR1300.html

Pennsylvania Department of Labor and Industry, "High-Priority Occupations (HPOs)," undated. As of October 9, 2014:
http://www.portal.state.pa.us/portal/server.pt?open=514&objID=814812&mode=2

Pittsburgh Regional Alliance, internal documentation provided to authors, 2013.

Power of 32, home page, undated (a). As of October 9, 2014:
http://www.powerof32.org/

———, "Sponsors/Partners," undated (b). As of October 9, 2014:
http://www.powerof32.org/about/?Sponsors-Partners-1

Smits, W., "Industry-Specific or Generic Skills? Conflicting Interests of Firms and Workers," *Labour Economics*, Vol. 14, No. 3, June 2007, pp. 653–663.

U.S. Energy Information Administration, "Profile Data," last updated September 18, 2014. As of October 9, 2014:
http://www.eia.gov/state/data.cfm?sid=PA

*Western Pennsylvania's Workforce Development System: Challenges and Opportunities,* undated. As of August 31, 2013:
http://www.trwib.org/admin/uploads/
Western-Pennsylvania-Workforce-Development-FINAL.pdf

World Coal Association, "Coal Mining and the Environment," undated; referenced May 10, 2013. As of October 8, 2014:
http://www.worldcoal.org/coal-the-environment/coal-mining-the-environment/